WAW

Choosing a Career in the Toy Industry

Working in the toy industry is a great career choice and can bring out the kid in you!

Choosing a Career in the Toy Industry

John Giacobello

The Rosen Publishing Group, Inc.
New York

Published in 2001 by The Rosen Publishing Group, Inc.
29 East 21st Street, New York, NY 10010

First Edition

Library of Congress Cataloging-in-Publication Data

Giacobello, John.
Choosing a career in the toy industry / By John Giacobello.
p. cm. — (World of work)
ISBN: 978-1-4358-8763-3
1. Toy industry—Vocational guidance—Juvenile literature.
[1. Toy industry—Vocational guidance. 2. Vocational guidance.] I. Title. II. Series.
TS2301.T7 G47 2001
688.7'023—dc21

2001001133

Manufactured in the United States of America

Contents

Introduction

At sixteen years old, Robert was like an overgrown kid—he loved toys. And even though people told him he was being childish, he never stopped collecting his favorite toys. He thought of them as treasures. Robert enjoyed taking things apart to see how they worked and then putting them back together again. His room was filled with model cars, electric trains, and games of all sorts. His friends even called him "Junior" and "Kid" because his hobbies were so unusual for his age.

Robert's parents appreciated his good nature and enthusiasm, yet they were also concerned for his future. "What's he going to do when he's older? He can't build toy trains and fly model airplanes for a living!" his father complained. His mother just laughed and said, "Maybe he will!"

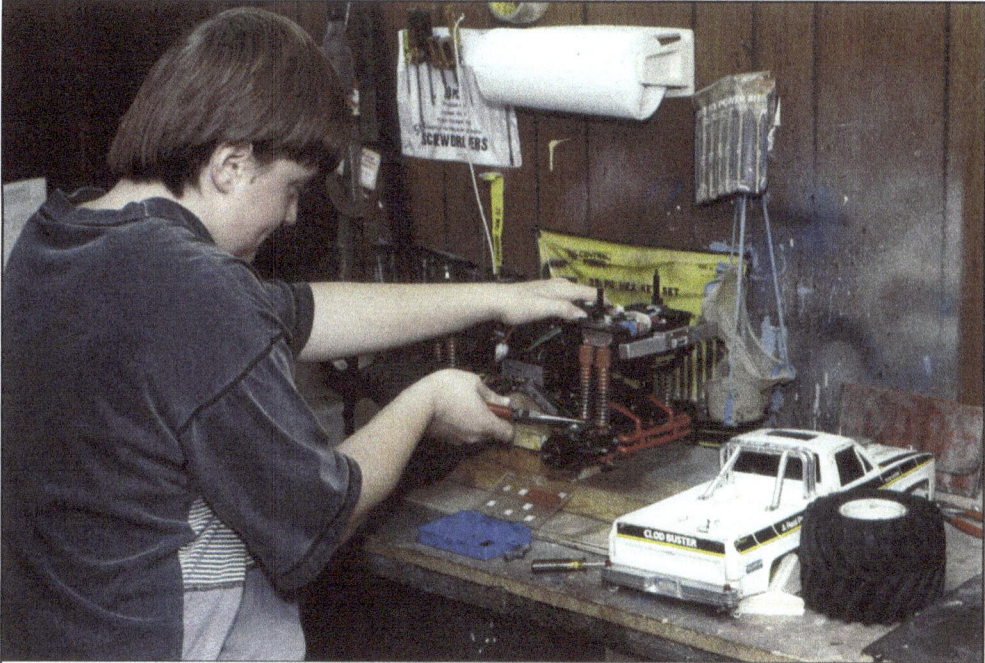
If you still love the toys you played with as a kid, designing toys may be the career for you.

Robert overheard them talking and started to worry. "What am I going to do for a living?" he wondered. When he spoke with the career counselor at school, she asked him about his interests. Robert reluctantly confessed his love of toys. "Well, that's a great start!" his counselor said. She showed him the Internet sites of major toy companies, such as Toys "R" Us and Hasbro. Many of them contained job listings! "Why have I never thought of this before?" he wondered. "I could make money doing something I love!"

Do you remember your favorite toy from childhood? That one item probably kept you

A career counselor can help you find a job that utilizes your interests and skills.

company, brightened rainy days, and entertained you during your most special years. Many adults have made fulfilling careers bringing the joy of toys to children all over the world.

The toy industry is an enormous business that takes in billions of dollars every year and is growing steadily. It is diverse, including everything from plush top-sellers like Furbies and Teletubbies, to action-packed video gaming systems such as the Sony PlayStation. The toy industry also produces dolls and action figures based on popular movies like *Toy Story*, as well as many other kinds of toys. And the kinds of toys being produced are constantly changing to meet the wants and needs of growing children of different backgrounds, ages, and education levels.

Because it takes many steps to bring a toy from a person's imagination to the public, there is a wide array of career possibilities in this field. Opportunities abound in areas such as design, accounting, marketing, and engineering. And beyond the specialties, there are still other areas of toy production to get involved in. If you have an interest in toys, there is probably a career in the industry that would suit you, no matter where your particular skills lie.

So if you still love the feel of a soft toy animal, or admire the design of a cool Matchbox car, have no shame! You just might be able to use your love of toys to build a bright future. This book will help you learn more about the opportunities available in this exciting field.

1

Welcome to Toyland

"*Come on!*" *shouted Tina, running with a large pink and black kite flapping behind her. Jennifer laughed and caught up to Tina. Jennifer was baby-sitting her niece today, and had decided to take her to the park to fly her new kite. Tina had been talking excitedly about the kite for a week now. Watching the little girl play, Jennifer realized how important this new toy was to her niece. "I hope the people who make and sell these toys understand how important their jobs are!" she thought.*

Tomorrow, Jennifer would become one of those people herself. She was starting a new job with a large toy company. Her supreme organizational skills and love of children had finally paid off, and she had landed a job as assistant to the marketing manager. Jennifer was nervous. What would they expect her to do? What kinds of issues would she have to deal with? What difficult lessons would she have to learn?

Tina shrieked with joy as the kite flew high in the air, and Jennifer smiled, thinking about the great opportunity that she had been given. She vowed to take her new career in the toy industry seriously and to think of Tina, and all the children playing with their different toys, every day when she went to work.

Toys in the Past

It is difficult to imagine the world without toys and games. Some say that people have probably created and played with toys since the beginning of time. We have found some playthings of ancient cultures, the most well-known example being the board game backgammon. This is a game that was invented 5,000 years ago. Today it is still popular and can be bought through any major toy seller. Chess is another game of the distant past that is still going strong. It originated in the sixth century in India, and then became popular in China.

You probably think of action figures as a modern invention. But the idea of small figurines that resemble people is extremely old. Okay, so no ancient artifacts have bendy arms and legs or removable jet packs and laser guns. But many figurines that have been unearthed seem to have been used for both entertainment and religious purposes. And the dolls favored by the kings of Europe did wear some elaborate and valuable outfits.

Monopoly's "Rich Uncle Pennybags" shows off a new Monopoly game token at the FAO Schwarz toy store in New York City.

Many games have a powerful connection with history that is not so ancient, but equally interesting. Monopoly, one of the most popular board games of our time, was invented in the United States during the Great Depression. During this period many people lost their jobs. One particular out-of-work man, Charles B. Darrow, sat in his kitchen, fantasizing about being a wealthy businessman. He turned his fantasies into a game, using his tablecloth as a playing board! He developed the game into the classic American pastime it is today.

Toys in the Present

The current toy industry is in a constant state of change. What's hot now may not be hot a year from now, or some new item may blow everything else away next month. Tickle Me Elmo was all the rage in 1997, Teletubbies took off in 1998, and 1999 was the year of Furby and Pokémon mania. Christmas of 2000 saw a surprising craze for robotic dogs! Kids' tastes are unpredictable, and toy companies are always scrambling to latch on to the hottest new trends. Every new toy is a risk, which keeps the industry fresh and exciting for many of its workers.

Playing Safely

Beyond questions of what will sell, toy makers have other issues to consider that you may not be aware of. Safety is a large concern. Nothing is worse for a toy company's reputation than producing toys that prove to be harmful to children. Manufacturers in any industry must earn customers' trust. But customers are much less trusting with their children's safety than with their own, so toy companies have to be certain that their products live up to every possible safety standard.

Toy makers must be especially careful with latex balloons, as children can swallow them and suffocate. Toys with small parts also can be deadly for young children, because small pieces are so easy for curious kids to swallow. And manufacturers must always use materials that are nontoxic. This is very important for products that are sometimes accidentally eaten, such as paint or paste. Great care must be taken that metal toys, like trucks and cars, never have sharp edges that could cut a child's skin. There are safety guidelines outlined by the government that all toy makers must conform to.

Think Like a Kid!

Toy makers must also become experts in child psychology. How else could they know what toys will hold a child's attention? Toys can elicit many different emotions from kids, such as excitement, curiosity, comfort, or even fear. The adults who create the toys need to understand the effects that their toys will have on a child's feelings. No parent will buy a stuffed animal that sends their son or daughter screaming from the room! And no kid will beg their parents for a board game that bores them to tears. Toys should be designed to bring out the appropriate feelings in children who play with them or the product is doomed to fail.

There is also a sense of responsibility that comes with producing toys. Kids learn a lot through play, so toy makers have a strong hand in shaping the way children come to view the world. This raises questions about toys and violence. There are many violent toys on the market, which often become very popular. Some toy companies say that they are not responsible for what these toys may be teaching children. And some parents do not mind their children playing with these toys.

But there is a growing feeling among many parents that toy makers should not market violent products to children. Organizations like The Lion and Lamb Project give these parents a way to organize and make their voices heard. They encourage parents to avoid buying violent toys, such as military, wrestling, or other combat-based items. Some wrestling games encourage things like

Giese Elementary School students turn in toy weapons before school in Racine, Wisconsin. They were rewarded with a hay ride and football-player trading cards.

"head smashing" or "arm twisting." These moves are intended for a fictional opponent, but parents worry that children learn through imitation.

The Lion and Lamb Project receives a great deal of media coverage for its efforts against violence in children's entertainment. Participating parents publish newsletters, organize events, and even testify at government hearings. Toy makers are forced to take these parents seriously, because bad publicity could hurt business.

On the positive side, many toy companies do produce excellent nonviolent toys that are fun and popular with kids. Some of these toys are even educational. There are trivia board games, art activity books, outer-space explorer kits, puzzles, and more that are entertaining to kids while also helping to sharpen their minds. Not all toys have to teach a lesson, but educational value can be an excellent selling point for a product. Most parents look for every possible opportunity to help teach their children. They might be more likely to buy a toy if it helps their kids learn while they play.

Toy Trends

Toy manufacturers need to keep up-to-date on trends in our society. For example, dinosaurs might be popular one summer. There could be a popular dinosaur movie in theaters, along with television specials about extinct animals, and dinosaur T-shirts and other products might be selling out in stores. This is the kind of marketing wave that toy companies must be able to detect early. Trend-related toys must hit the shelves before the craze reaches its peak. If they catch on too late, the trend will be finished by the time the toys become available.

Technological trends and advances are also essential to the toy industry. Toys have always utilized cutting-edge technology to attract customers. Look at the progression of video games: At one time, the best games you could get at home had simple graphics and made monotonous beeping sounds. Now home gaming systems provide digital graphics that look like Hollywood movies, with sounds and music to rival the hottest MTV videos. Sometimes an entire toy can be based on a new technology. There was a time when Etch A Sketch, the classic magnetic drawing board, was a new technology. The same goes for a sticky octopus that can climb down a wall, or a foam Nerf ball you can throw at your friends.

The newest technologies are often utilized to make new toys.

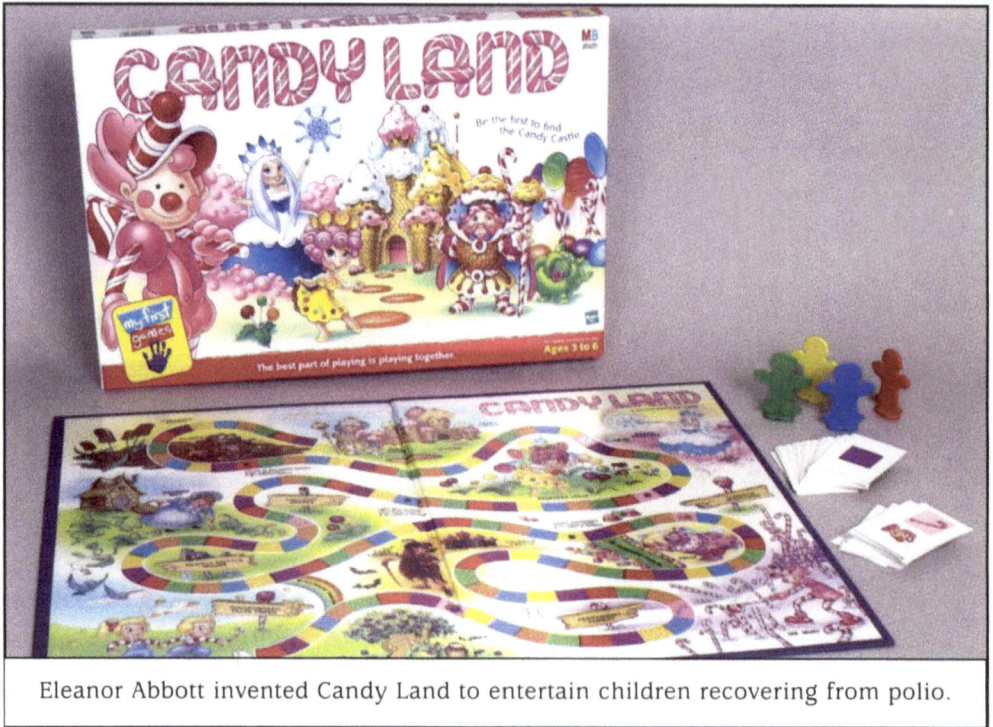

Eleanor Abbott invented Candy Land to entertain children recovering from polio.

How Did That Happen?

Have you ever wondered who invented your favorite toys, when, and why? Every toy has its own unique history, and the history and evolution of some toys are unusual stories. Here are some fun facts about the toy industry you probably did not already know.

Candy Land

Eleanor Abbott of San Diego invented the popular board game Candy Land, known internationally as "a child's first game." It was the 1940s, and Eleanor was recovering from polio. She was surrounded by children with the disease, so to keep herself busy she would come up with activities to entertain the kids. Candy Land was such a major hit with the kids, she decided to submit the idea to the Milton

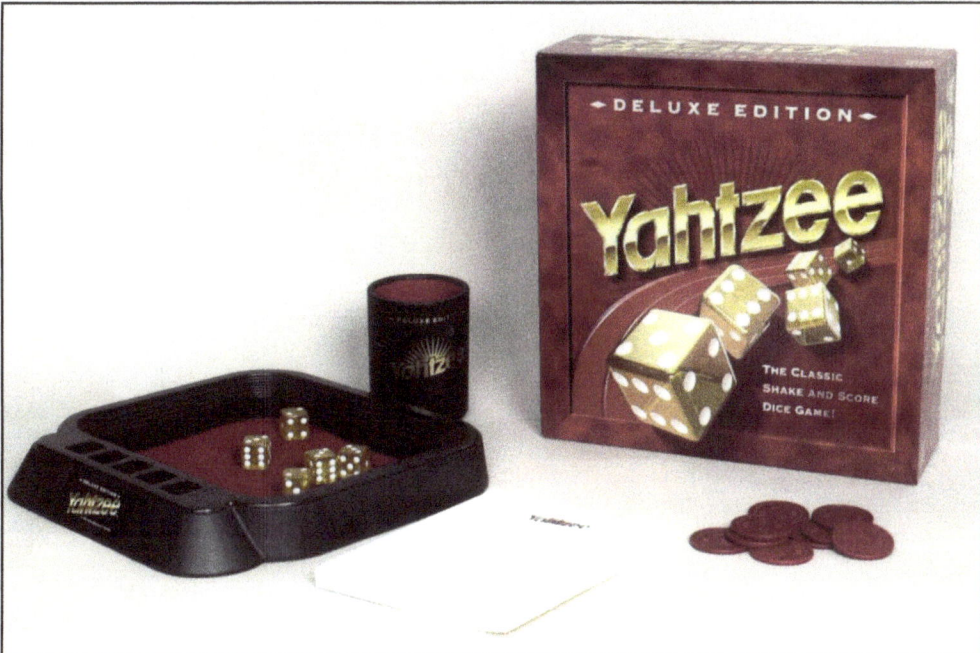

The popular dice game Yahtzee was created by a couple who owned a yacht.

Bradley Company. It accepted, and Candy Land went on to become one of the company's most popular items.

Yahtzee!

Where did this bizarre word come from? The popular dice game Yahtzee was created in 1956 by a Canadian couple who owned a yacht. They would always play the game with their friends while cruising on the yacht, so they began calling it their "yacht game." Their friends enjoyed it so much, the creative couple wanted to have a few of the games duplicated to give away as gifts. They presented it to game manufacturer Edwin S. Lowe. He loved the yacht game and offered to buy the rights. The couple was only interested in the few copies for their friends, so they let Lowe have the rights for free. Yahtzee was a smash and was eventually bought by Milton Bradley in 1973.

The Koosh Ball

Have you ever played with a Koosh Ball? It's a small ball with soft, rubbery strands sticking out all around it. The toy somewhat resembles a porcupine. An engineer named Scott Stillinger created the Koosh Ball in 1987. His children had been having trouble catching a regular ball when it was thrown to them, because their hands were too small. So he created a ball made of rubber bands that would be easier to grab. He named it "koosh" after the sound it made when it landed in his hands.

2

Careers Making Toys

Before we jump into the complex world of marketing and selling toys to the public, let's first look at how products are created. Talented inventors, designers, engineers, painters, and inspectors are all necessary to create a new toy. Are you an idea person? Do you like to work with your hands? Do you have artistic skills? In this section we will discuss the beginning of a toy's life cycle, and the people who make it possible.

Toy Inventor

Ideas are extremely valuable in all industries. Very few people have the talent to come up with great new ideas for products. So those who show an ability to produce exciting, unique, and well thought-out concepts for new toys will always be in high demand.

Toy companies need qualified employees to work behind the scenes.

Janet Lyons practices on the Pooh Counting Carrots toy. Lyons will demonstrate the toy at the Tiger Toys exhibit during the annual American International Toy Fair in New York.

But what does a toy inventor need to consider when coming up with a new idea? First of all, the toy has to be something that is likely to sell. Although nobody can tell for sure what will or will not sell before it hits the market, the inventor must be in touch with children enough to understand what kids enjoy. The toy should also have staying power. Word gets around about toys that become boring after a few hours of play. And the idea needs to be cost effective. No company will be interested in a toy that requires high-grade metal and rare, expensive chemicals to manufacture.

Some inventors work independently. This means that they do not work for one company alone, but freelance by selling their ideas to many different toy manufacturers. These independent inventors have the most success with small- to medium-sized companies, since the large manufacturers tend to generate ideas from within the company.

If you decide to pitch an idea, be prepared to accept rejection. Your ideas may be turned down many times before being accepted, if they are accepted at all. If a company decides to manufacture your toy, you will probably be paid a royalty. That means you earn a percentage of the toy's sales, the average toy royalty being 5 percent. How much you earn as a freelance toy inventor will vary depending on how many ideas you sell and how well the toys perform on the market.

For a more stable income, you could invent for a design firm or toy manufacturer full-time. These types of jobs usually require design skills, as well as great ideas. We will discuss toy design and the skills it demands in the next section.

Toy Designer

The toy inventor and the toy designer may often be the same person, but invention and design are two different tasks. As we discussed earlier, inventing the toy means coming up with the original idea. Designing it means making important decisions such as what the toy will look like and how it will function. Designers need to have artistic ability to communicate those decisions. A toy designer may earn between $30,000 and $60,000 yearly, depending on his or her skills and experience.

The toy designer often makes sketches of his or her ideas on paper, taking care to show the shape, texture, and color of the proposed toys.

The designer sketches the idea on paper, utilizing drawing skills to show qualities such as shape, texture, and color. Sometimes he or she will also use clay to sculpt a model of the toy. This task may also be done by a separate artist known as a sculptural designer.

Toy designers also determine the age group a toy would be best suited for, and figure out about how much it will cost to manufacture. If the product idea is a game, he or she must write the game's rules. If the product idea is a toy, the designer needs to provide instructions for its use.

A designer may work on many toy ideas that are never manufactured. The company may see the designs and decide the toy is not worth creating. These ideas are then trashed. Designers must accept that much of their work will not be used, just as inventors must get used to the fact that many of their ideas will be rejected. Only the best of many ideas and designs will ever be turned into toys.

Drafting Engineer

Drafting is a kind of engineering. Toy drafting is similar to toy design, in that it requires excellent drawing skills. The drafting engineer takes a design one step further by creating an extremely detailed drawing, using the designer's original sketches or clay sculpture. Most drafters earn salaries between $25,000 and $40,000 per year.

This detailed drawing shows how all of the pieces of the toy would fit together, and how even the tiniest parts function in the design. The

The drafting engineer creates a detailed drawing of the toy using the designer's original sketches or clay sculpture.

drawing is then used as a blueprint to create the toy. Today, most drafters use computers to help create the blueprint. Drafters also specify the best method for manufacturing the toy, and what kind of materials should be used. If a designer is skilled in engineering, he or she may perform the drafting. But most of the time, a company will employ engineers to draft the details of the designer's work.

Drafting requires great precision and attention to detail. Think about a dollhouse with doors that open and close. What if the doors were made the wrong size? Or what about a toy spacecraft that has detachable rockets on each side? How do the rockets drop off when you push a button, and then easily snap back into place? You can imagine how complicated many toy designs can become.

Molding, Diemaking, and Painting

Plastic toys and their parts are manufactured using molds. A mold is a piece of steel that is formed into the shape of a toy, or one of the toy's pieces. When melted plastic is poured into the mold, it cools and hardens in the mold's shape. These molds fit into machines used to manufacture an endless supply of plastic toys and parts.

Mold makers are the people who use the drafter's blueprints to shape molds from steel. Accuracy is just as important to a molder as it is to a draftsman. If a molder makes even the tiniest mistake, thousands of toys could be manufactured incorrectly. This error could cost a company a great deal of money and time to correct.

But what about toys that are made out of metal instead of plastic, like Matchbox cars and Tonka trucks? Metal toys are made using special tools called dies. While soft plastic can be poured into molds, hard metal must be cut, bent, and stamped using dies. A new die must be created for every toy design, so skilled diemakers are always in demand.

Some toys and parts need to be painted. Machines handle this task for the most part, but workers called paint sprayers are still required. These workers operate the paint machines and spray paint any spots missed by the machines. They use adjustable spray guns to control the amount of paint that is applied to the toys.

Mold and die makers may earn from $20,000 to $50,000 yearly, while paint sprayers and paint machine operators can earn between $20,000 and $30,000 annually.

Quality Control

After a new toy has been invented, designed, and engineered, there are still more steps to go through before a toy is ready to be sold in stores.

First, a toy must be tested to make sure that it is safe enough for children to play with. In chapter 1, we talked about how important toy safety is to the reputation of a company. Guaranteeing the safety of all toys and games produced is the vital responsibility of the quality control manager.

The quality control manager performs a variety of tests on the company's products. The tests are designed to ensure that the toys are able to meet the safety standards outlined by the United States government. The government agency that sets these standards is called the U.S. Consumer Product Safety Commission (CPSC).

Quality control includes not only the safety of the toy, but its durability as well. The manager may throw toys, crush them with tools, or drop them repeatedly. These product testers and inspectors may earn $20,000 to $32,000 per year. They help the company find out how much rough play a toy can stand up to, and how long it will last in the hands of an energetic child.

How to Prepare for a Career Making Toys

If you are interested in a career making toys, there are many ways to prepare. A college degree is not always necessary, but you may require some training beyond high school in order to develop

29

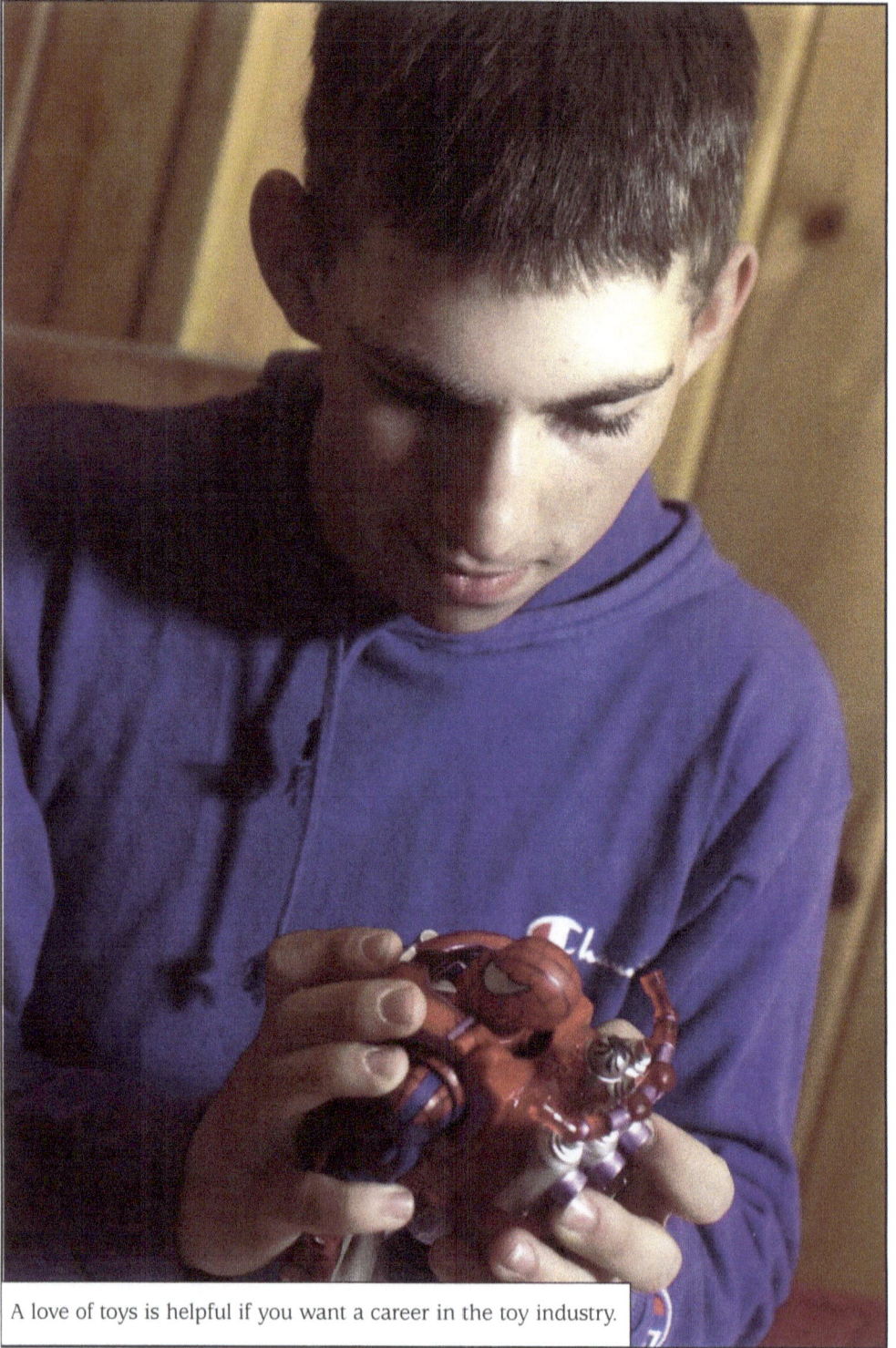

A love of toys is helpful if you want a career in the toy industry.

the necessary skills. The first school to offer a comprehensive toy design degree was the Fashion Institute of Technology (FIT) in New York City. Some graduates of this program went on to develop successful toys and games such as Tickle Me Elmo, Hot Wheels, and Don't Spill The Beans. Almost all of the students who complete the program are hired by toy companies. This program is excellent preparation for any of the careers described above, as its classes cover just about every aspect of toy production.

Some other schools featuring toy design programs are listed at the back of this book. Other excellent college and technical school coursework includes graphic design, drawing, sculpting, drafting, and child psychology.

You can also begin to prepare simply by looking closely at toys. Try taking them apart to see how they are designed, and think about how all of the pieces fit together. If you have access to the Internet, perform a search for different toy manufacturers. Looking over the Web sites of toy makers will give you a great overview of the companies you could work for and the products you may someday help to create.

3

Careers Selling Toys

*J*im had started working as a cashier at a Toys "R" Us during the summer between his junior and senior year in high school. He liked the job because he enjoyed working with people, and the income helped him to save money to buy a car. Jim handled an incredible number of toys each day, and his eyes were always drawn to the bright colors and slick designs of the packaging. He had taken a few drawing classes at school, so he could really appreciate the skill it must have taken to get the designs right.

One day while stocking shelves, Jim stopped working to take a close look at the incredible array of graphics on the boxes. Pretty soon he was doodling these graphics on pads of paper while on his break, and creating designs on his computer at home. "How cool would it be to design the packaging for

Working part-time at a toy store can help you learn about the toy market if you are interested in pursuing a career in the toy industry.

toys?" he thought. "You would get paid to draw monsters, color superheroes, and come up with cool patterns and prints." He wondered how he could get a job that was so much fun.

If you are not interested in a career making toys, maybe you could explore the business side of the toy industry. Skilled artists are needed to create the eye-popping graphics that make toys jump off shelves and into shopping carts. But there are also opportunities for thorough market researchers, persuasive salespeople, and organized managers. If you might be interested in a serious business career within a playful industry, read on.

Companies often use children to test and rank toys to try to figure out which toys will be big sellers and which will be duds.

Market Research

There is always an element of risk when a company produces a new toy. Nobody really knows if the product will succeed or fail. But companies can reduce that risk by conducting market research.

Sometimes this can be as simple as inviting children to come play with the new toy. Market researchers watch the kids' reactions and ask questions. Did they enjoy the product? Were they bored? Which age group liked it the most? Were they confused by the instructions? There is a good chance that if the toy or game went over well with several random groups of children, it will succeed on the market.

Market researchers also study statistical data. This means that they look at numbers of different types of toys sold in different time periods, in various cities and states, and to different groups of people. They may analyze toy sales based on customers' family size, income level, race, and other factors. This helps the company to predict who will be most interested in a new toy, and what groups of people it would be best to market to.

A market research analyst can earn anywhere from a starting salary of $35,000 to around $75,000 for experienced market research executives. This is an excellent career for anyone who is interested in other people. A market researcher in the toy industry should have a natural curiosity about what makes kids feel the way they do, and some basic knowledge of child psychology.

Marketing Services

While market researchers dig up information on which groups of people to sell to, the marketing services director uses that information to come up with the best ways to market the product. The major way this is done is through advertising campaigns. You see commercials on television all the time. Other types of advertising include magazine ads, billboards, Internet banners, and whatever else the marketing department can dream up.

It is up to the marketing services director to decide the best method of advertising to reach the right groups of kids. For example, an action game designed for boys between the ages of seven and twelve might be best advertised through a television commercial on Saturday morning. This would get the attention of the "cartoon crowd." Or a new doll could

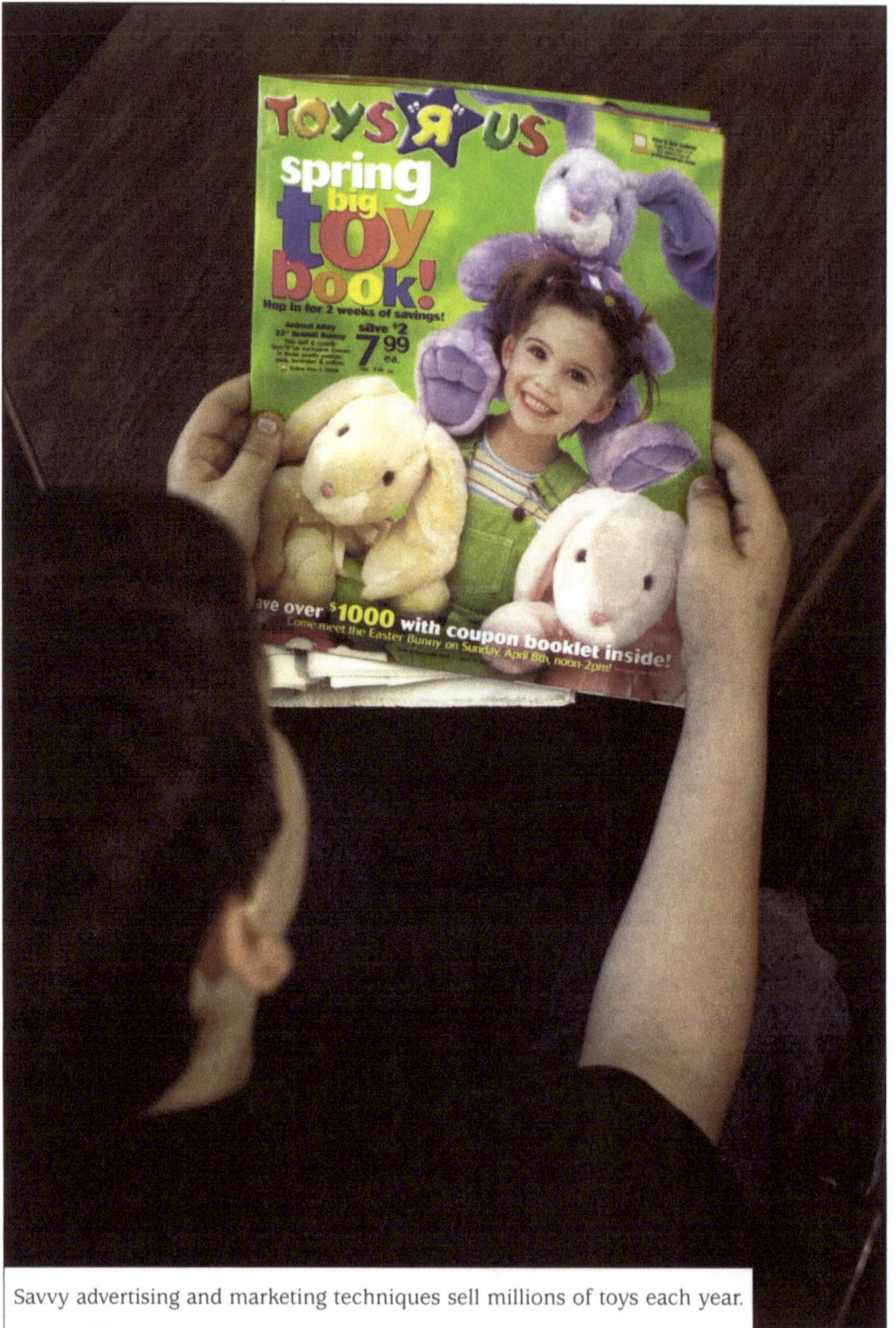

Savvy advertising and marketing techniques sell millions of toys each year.

be introduced through a young girls' magazine such as *American Girl* or *Girls' Life*. Marketing services may come up with all sorts of interesting ideas to promote the product, from contests and giveaways to direct mail campaigns and movie tie-ins.

The marketing services department may have a staff of people to write and design the advertising materials. Or they may hire an outside agency to create the advertisements necessary to launch a campaign. Those who do the writing for advertisements are called copywriters or editors, and those who create the visual artwork are called art directors. Marketing services also include putting together the company's catalog. The catalog is an excellent advertisement in itself, in that it presents every product available through the company along with its price and description. An exciting catalog can have a positive effect on a toy company's sales.

Marketing assistants generally earn between $30,000 and $43,000 per year, while marketing managers may earn as high as $85,000.

Packaging Design

Think about all the different products you see every day. How do you tell them apart? The one thing that sets products apart from one another at first glance is the packaging. And when you are scanning the shelves of a store, that first glance is extremely important. Each product is sold in a unique type of package, with its own shape, color scheme, and style. Creating this style is the job of the packaging designer.

Toy makers rely on packaging and design to set their toys apart from others so potential buyers will notice them on crowded toy store shelves.

Packaging design is the perfect opportunity for someone who possesses both artistic vision and business savvy. A packaging designer has many responsibilities, and usually earns between $28,000 and $40,000. He or she must be able to create packaging for toys that is not only visually appealing, but that will make kids and parents want to buy the toys as well. The packaging should also be able to convey what the toy inside is like. Fun, exciting toys should have explosive packaging to capture the spirit of the product. Soft, comforting toys should use more subdued colors and a subtle design to suggest the idea of comfort.

Products also have their own logos, which are easy to recognize. A logo is just the name of the product, written out in a special way. For example, think about the popular toy sensation Pokémon. When you see the word Pokémon, it is usually written in big yellow letters outlined in blue, and the word arches like a rainbow. That is the Pokémon logo. The packaging designers of Pokémon were also clever enough to add a sentence to the logo, which usually appears beneath the product name. The sentence reads, "Gotta catch 'em all!" This makes kids excited about collecting all of the Pokémon products that are available.

Toy Sales

After a toy is manufactured, it has to be bought—first by a toy store and then by a customer. To accomplish the first step, the manufacturer employs a sales manager to show the product to toy stores. Sales managers, and often a staff of salespeople, will travel to the stores and demonstrate the products in person. These workers need to have excellent people skills, since a big part of the job is building trusting relationships with the store's buyers.

A buyer works for a toy store and decides which items the store will stock. Buyers have to stay on the cutting edge of toy industry trends in order to have an idea of which toys customers will want most. Buyers also consider the quality and price of the products before making a decision that will best benefit the toy store.

The workers who interact directly with the customers are called retail salespeople and managers. These workers should also deal well with other people, and play close attention to detail. Retail sales careers can be extremely demanding. Salespeople may have to work late hours, and are expected to learn everything there is to know about the products they sell. Customers may have many questions about the products being sold. If a salesperson does not know the answer to a question, he or she must be willing and able to find out the answer. Salespeople have to work hard to satisfy customers so they keep returning to the store. Anything the salesperson cannot handle is turned over to his or her manager, who is generally the more experienced worker. The manager oversees the work of the salesperson.

How to Prepare for a Career Selling Toys

A career as a packaging designer or advertising art director may require some specialized art training. Most art schools that offer two- to four-year training programs should be able to provide you with the necessary skills to perform these jobs. The other careers in this chapter do not necessarily require a college degree, but some sort of schooling beyond a high school diploma is extremely helpful when competing in today's intense job market. Courses in professional and creative writing, business, economics, and marketing would be excellent preparation for this field.

The most important thing is to have an understanding of what kids want. Try to observe children playing. What do they enjoy doing most? What seems to make them happy, and what makes them walk away or cry? Talk to younger brothers, sisters, or other relatives about their favorite toys.

Try starting at the bottom. If you are looking for a job while still in school, accept a position as a cashier at a toy store. While performing that job the very best you can, keep your eyes and ears open and learn everything you can about what is happening around you. You may be promoted to manager if you perform well. Then a buyer may need an assistant. Eventually, you may be able to work your way up to buyer.

The same career ladders exist with toy manufacturing companies. Someone may start out answering phones and fetching coffee, and end up leading the marketing department. Anything can happen if you work hard and pay attention. And when you see an opportunity opening up, do not hesitate to go for it!

4

Other Career Possibilities

The toy industry has become so diverse and specialized that many career opportunities have opened up in the past decade that may not have existed before. The Internet has become a real buying tool for consumers from all walks of life. Video games and gaming systems have become so technologically advanced, and so popular, that they are an industry in themselves. And the unique needs of children with disabilities are being recognized by a growing niche within the toy industry.

Toys and the Internet

Jake considered himself a pretty old-fashioned guy. He did not even own a computer until his daughter bought him one for his birthday. He hardly ever used the thing. It was just so confusing! There were all of these things to click and stuff popping up out of nowhere. He had decided to stay off of the information superhighway for good.

But the holidays were coming. The thought of dragging himself out to those crazy stores, and fighting old ladies over the last Barbie doll, just made him sick. But he wanted to buy some great gifts for his grandkids. Jake had heard that some people did all of their holiday shopping using the Internet. "Well, maybe I could try one more time," he thought.

With a little help from his daughter, Jake found the items he wanted and placed orders using his credit card. His shopping was finished in about an hour! "That was so easy!" he exclaimed. "I don't know why I avoided the Internet for so long. What a helpful tool!"

More and more people today are shopping from home, buying products over the Internet. They can avoid crowds, especially during the holiday rush, and select their purchases while sitting in a comfortable chair at their computer.

Electronic shopping can also be more organized than scanning the aisles at a toy store. There are many Web sites that enable shoppers to find toys based on the age and gender of the child they are shopping for. Shoppers can also search by a toy category, such as puzzles, dolls, and board games, or by toy brand. Once the products are chosen, the shopper pays for the items by entering a credit card number. The toys are mailed to the address provided by the user.

More people are turning on to this type of shopping experience each year. Online retailers include SmarterKids.com, Sanrio.com (for all of those adorable Hello Kitty products!), ToysRUs.com, and many more.

It may seem as though the electronic shopping craze would eliminate retail jobs in the toy industry. What happens to all of those cashiers, salespeople, and retail sales managers if people are using virtual shopping carts instead of real ones? True, no Internet toy company has any use for retail salespeople. But exciting new possibilities in technology careers have been created. These companies need workers to develop and maintain Web sites, oversee computer networks, and help process information.

If you have a knack for computers, and also have an interest in toys and games, the electronic toy-shopping realm may be an excellent place for you to work. Try to take as many computer courses as possible, and experiment with designing your own Web sites. There are also many books available on programming languages like HTML, which is used in Web design. Companies hiring in this field are not necessarily looking for college degrees. But you do need to know your way around a computer to break into the field, even as an assistant. If you decide you need some training, many colleges and technical schools offer courses in Web design, programming, and networking technologies. These skills can also be learned at home through courses offered over the Internet.

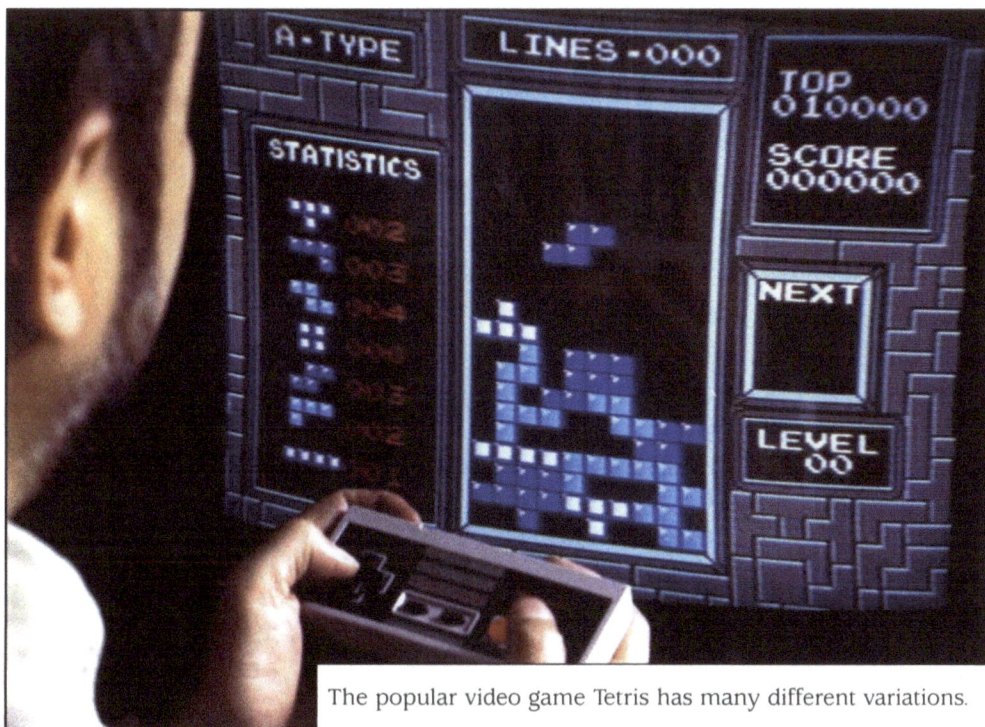
The popular video game Tetris has many different variations.

Video Game Design

Susan nudged the joystick, narrowly escaping death from an atomic death ray. "Same old game," she thought. "Why can't these companies come up with an original idea?"

Susan had a lot of ideas for fun video games. She sketched out characters, wrote stories, and even imagined music to go with her ideas. The games came to life in piles of notebooks. Susan crafted complex video adventures complete with difficulty levels, scenery, and animation. And she took pride in the fact that her games were totally cool without being violent.

These were the kinds of games Susan wished she could buy for herself. She had heard about software you could use to create your own video games, and had thought about trying it. "It would be so cool to actually play all of these games I've dreamed up," she told herself. "Maybe I could even sell my ideas and make a little money."

You do not have to be a computer whiz to design a video game. There is user-friendly software available that can help you translate ideas into computer programs, without having to learn complicated programming languages. This software includes AGAST (Adventure Game Authoring System), Stagecast Creator, and HyperStudio.

Some video game designers work with computer programmers to develop their ideas. Designers may use sketches they have drawn, scripts they have written to describe the object and levels of the game, and any other ways they can think of to communicate their ideas to the programmers. The programmer takes this information and turns it into a language that the computer can understand.

Animators are needed to create the artwork for the game. The artwork may be drawn on paper and scanned into the computer, or created using a computer drawing program. Musicians create the soundtrack and other sounds for the game. Testers try out the games as they are developed and help the programmer to find problems. A problem with the game may be anything from action that is too slow and boring, to technical details like characters that do not move properly. This can be a fun job, but it is not as easy as it sounds. Video game testers have to pay close attention to tiny details.

Anyone can design a video game, as long as they have good ideas that they can communicate well. Drawing and programming skills are helpful, but not required. Good writing skills are necessary for explaining various aspects of the game, so pay attention in English class! There are also some colleges and design schools that offer coursework in video game design. No special education is required to test video games, but programmers, animators, and musicians all require special skills that may be learned through training or independent study. A love of video games helps, too!

Toys for Kids with Disabilities

Not all children have the same needs when it comes to toys. Parents of children with physical handicaps have a challenging job ahead of them when they go to pick out toys for their young ones. Shopping at a regular toy store can be difficult. They have to choose from toys that were designed for the average child. These parents must carefully select items that will entertain and stimulate children with hearing loss, visual impairment, and other disabilities.

Fortunately, there are other options. Some companies specialize in toys designed with the special needs of these children in mind. Toys can be therapeutic. That is, they can help kids learn to

Florence Henderson donates her time to help children with disabilities. These special toys give disabled children something with which to identify.

deal with their impaired senses. For example, a toy that moves across the floor while playing music is an excellent aid for visually impaired children in finding things by sound. Children recovering from surgery, or suffering from problems with their arms or hands, can play with brightly colored putty and other toys to strengthen their grip. Children with speech impediments can play games to help them pronounce words properly. And the best part is that kids have fun while getting the help they need. Companies such as the Dragonfly Toy Company, Kapable Kids, and TFH (Touch, Feel, Heal) make it easy for parents and therapists to find these useful toys.

It takes intensive training in highly specialized skills to design toys for special needs children. If you think this career may be for you, try to find schools specializing in fields like rehabilitation engineering, assistive technologies, and developmental play therapy. Rehabilitation engineering and assistive technologies both refer to the use of technology to help people with disabilities confront everyday difficulties. Play therapy is a way of using play to help children grow, develop, and resolve problems in their lives. Designing toys for the disabled could be a creative and rewarding career for the right person, who has the necessary skills.

5

Toy Stories: Real-Life Career Experiences

The best way to learn about any career is by talking to people in the business. If you are fortunate enough to find someone working in the toy industry who is willing to talk to you about his or her job, ask lots of questions! In this chapter, you can read what some people in the industry are saying about the careers they have chosen. What you hear may surprise you.

A Career as a Video Game Tester

Testing video games may sound like the ideal job, especially if you are a gaming enthusiast. Go to work, kick back, and play games all day long. Could it be heaven?

Not so, says Nancy Hardwich, a game tester. She explains that, believe it or not, this job can come with high pressure. The tester is the person who decides whether or not the game is suitable to be sent out to stores. If bugs slip through, the company looks bad and will probably lose sales. The tester is responsible for making sure this does not happen.

Nancy also describes how dull testing games can become after a while. She often has to play the same game, over and over again, for as long as four months or more. And the hours can be difficult. Many testers have occasional seventy-to-eighty hour workweeks, when they actually sleep in the same room where they work!

A Career as a Toy Designer

Jacob Coleman has been involved with designing some pretty important toys. After years in the business, he and some other experienced toy industry workers have started their own company. Jacob is an excellent example of a toy success story.

Before beginning his toy career, Jacob worked for a company that designed tools and medical equipment. He liked to draw cars in his spare time. He knew a woman whose husband worked for a popular toy company, and they needed someone to sketch cars for a market research project. They liked his work and offered him other freelance jobs. Soon he was offered a full-time job as a toy designer, and he accepted. Jacob believes that anyone going into toy design should be aware of everything that happens in popular culture. He also recommends art school, for training in drawing and sculpting, as well as courses in engineering and material processing.

A Career in Toy Market Research

Andrew Petersen is another example of an employee at a big toy company who went on to start his own company. Today Andrew is the president and owner of a company specializing in toy market research.

Andrew describes market research as an attempt at seeing into the future, since the testing he does takes place six months to two years before the product would hit store shelves. Not only do you need to know what kids want, but what kids are going to want next as well.

Andrew feels that a degree in business can be helpful to those who want to enter the field of toy marketing, but it's not necessary. He rates a love of toys as the most important qualification, along with excellent math skills, precision, attention to detail, and strong written and verbal communication skills.

A Career in Toy Advertising

Dean Potter works as an editor for a toy company. Dean is a college graduate who majored in English, and he says he took the job simply because it sounded like fun. He started out doing small freelance projects for the company. When a full-time position opened up, it was offered to him. His responsibilities include coming up with names for action figures and their weapons, as well as creating their biographies.

Potter says that in order to do this sort of writing, he has to try thinking like a child would. Dean says that creativity is the most important asset for a position like his, as well as an ability to work on a team. He describes his office environment as playful, complete with midday Nerf toy fights!

Steve Wozniak, creator of Apple's first computer, left, and Nolan Bushnell, founder of Atari Inc., have joined forces to build electronic children's toys.

Up the Toy Industry Ladder

One of the most fascinating stories of toy industry success comes from Tom Whitland. Tom started out as an engineering draftsperson. He helped to engineer a line of very popular metal toy car collectibles. Tom eventually worked his way up to being the marketing manager for the entire line of toy cars.

In between, Whitland did just about everything there is to do at a toy company. He invented and designed toys, seeing them through right up to their production. He spent some time doing research and development, and then quickly rose to the top of the marketing department. Tom is thankful he had the opportunity to do so many different types of jobs, and to gain so much valuable experience.

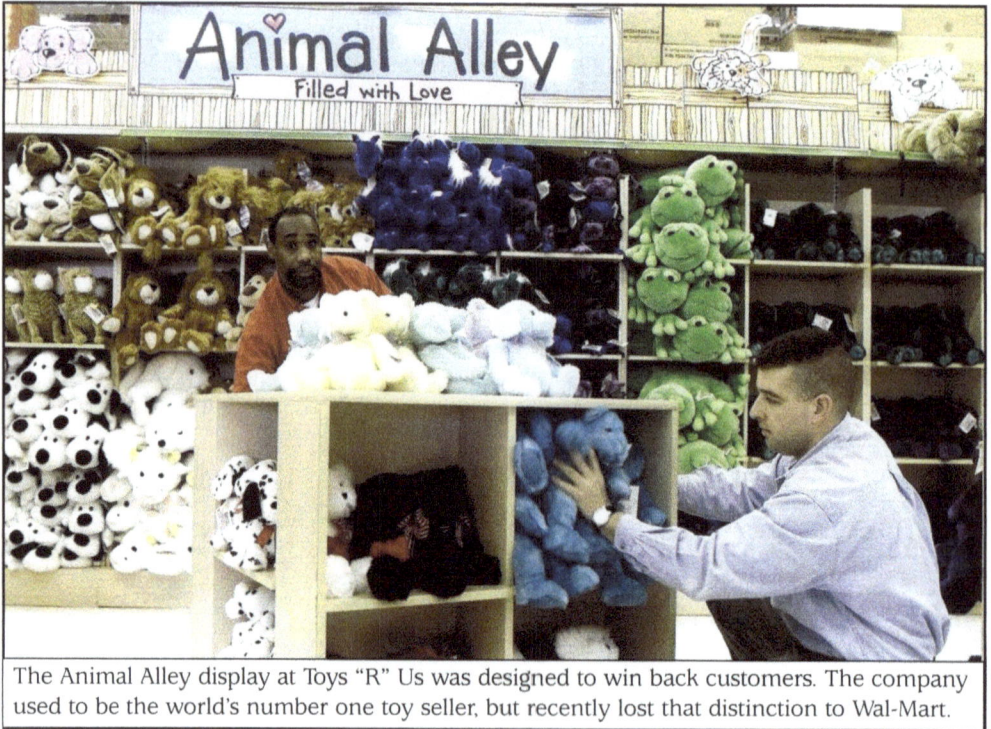

The Animal Alley display at Toys "R" Us was designed to win back customers. The company used to be the world's number one toy seller, but recently lost that distinction to Wal-Mart.

Eventually, Tom spent several years inventing toys. Fascinated with all things technical, he soon turned his attention to electronic games. He spent two years learning the high-tech toy business. Finally, Whitland was able to apply everything he had learned to running his own company.

From Idea to Product

In this book, you have learned about the many phases a toy goes through before it finds a home with a little boy or girl. It all starts with a brilliant idea in some creative thinker's mind. The idea is evaluated by researchers who are experts in understanding what kids want. Then the toy is designed and engineered by a team of skilled artists and craftsman. The toy is reproduced on a mass

54

scale by factory workers. Word goes out to the public that a fantastic new toy will soon be on the market, and kids hear about how cool it is. Finally, the toy is stocked on toy store shelves and sold by retail sales professionals.

Throughout this entire process, the toy passes through many capable hands. And all of these creative, intelligent people bring unique skills and knowledge to the success of the product. Whatever your strengths, you could apply them successfully in this fun, diverse, and dynamic industry.

Glossary

art director   Creates visual designs for
  advertising campaigns.
assistive technologies   The application of
  technology and scientific principles to meet the
  needs of, and address the barriers confronted
  by, individuals with disabilities. Also referred to
  as rehabilitation engineering.
buyer   Decides what products a toy store will
  stock; works closely with sales managers.
copywriter   Performs writing tasks for
  advertising campaigns.
developmental play therapy   Utilization of play
  in growth and development.
die   Tool used to cut, bend, and stamp metal to
  mass-produce toys or toy parts.
freelance   Working on projects for many different
  companies, rather than working full-time for
  one company only.
HTML   Computer language used to design
  Web sites.
marketing services director   The person who
  determines the best ways to advertise a product.

market research Methods of predicting the success of a product.

mold Piece of steel formed into the shape of a toy or toy part, used to mass-produce plastic items.

packaging designer The person who creates designs for toy packages.

paint sprayer The person who operates paint spraying machines and paints areas missed by the machines.

quality control manager The person who tests toys to ensure safety and durability.

rehabilitation engineering The application of technology and scientific principles to meet the needs of, and address the barriers confronted by, individuals with disabilities. Also referred to as assistive technologies.

royalty A percentage of an item's sales, usually earned by a toy's inventor.

sales manager Works with buyers to sell products to toy stores.

toy designer The person who decides what a toy will look like and how it will function.

toy drafter Engineer who creates a detailed drawing, or blueprint, for a toy.

toy inventor The person who generates ideas for new toy products.

U.S. Consumer Product Safety Commission Government agency that sets the safety standards that toy companies must meet.

For More Information

In the United States

Children's Advertising Review Unit
National Advertising Division
Council of Better Business Bureaus
4200 Wilson Boulevard, Suite 800
Arlington, VA 22203-1838
(703) 276-0100
Web site: http://www.caru.org

Fashion Institute of Technology (FIT)
Seventh Avenue at 27th Street
New York, NY 10001-5992
(212) 217-7999
Web site: http://www.fitnyc.suny.edu/academic/
all_majo/2.11.02.html

The Lion & Lamb Project
4300 Montgomery Avenue, Suite 104
Bethesda, MD 20814
(301) 654-3091
Web site: http://www.lionlamb.org

Otis College of Art and Design
9045 Lincoln Boulevard
Los Angeles, CA 90045
(800) 527-OTIS (6847)
e-mail: otisart@otisart.edu
Web site: http://www.otisart.edu/programs/BFA/
    Toy%20Design/toy.htm

Toy Manufacturers Of America, Inc.
1115 Broadway, Suite 400
New York, NY 10010
Web site: http://www.toy-tma.com/index.html

In Canada

Canadian Toy Association
P.O. Box 294
10435 Islington Avenue
Kleinburg, ON L0J 1C0
(905) 893-1689
Web site: http://www.cdntoyassn.com

Canadian Toy Testing Council
22 Antares Drive, Suite 102
Nepean, ON K2E 7Z6
(613) 228-3155
Web site: http://www.toy-testing.org

Dragonfly Toys (for kids with special needs)
291 Yale Avenue
Winnipeg, MB R3M 0L4
(800) 308-2208
Web site: http://www.dragonflytoys.com

Web Sites

PlayDate Inc.
http://www.playdateinc.com/industry/
    toyindustry.htm

Toy Tips—The Independent Voice on Toys for
    Child Development
http://www.toytips.com

For Further Reading

Cross, Gary. *Kids' Stuff: Toys and the Changing World of American Childhood.* Cambridge, MA: Harvard University Press, 1997.

Lerner, Mark. *Careers in Toy Making.* Minneapolis, MN: Lerner Publications Co., 1980.

Lund, Bill. *Getting Ready for a Career as A Video Game Designer.* Mankato, MN: Capstone Press, 1998.

Oppenheim, Joanne F. *Buy Me! Buy Me! The Bank Street Guide to Choosing Toys for Children.* New York: Pantheon Books, 1987.

Stern, Sydney. *Toyland: The High-Stakes Game of the Toy Industry.* Chicago: Contemporary Books, 1990.

Index

About the Author

John Giacobello is a freelance writer living in New York City. He is also the author of *Choosing a Career in Music*, *Exploring Careers in the Fashion Industry*, and *Scuba Diving: Life Under Water*.

Photo Credits

Cover by Cindy Reiman. P. 2 © Corbis; pp. 7, 27 © Image Works; pp. 8, 18, 19, 30, 36 © Cindy Reiman; pp. 12, 15, 22, 23, 33, 34, 38, 45, 48, 53, 54 © Associated Press; p. 17 © Toy Manufactures of America; p. 18 CANDY LAND® & © 2001 Hasbro, Inc. Used with permission; p. 19 YAHTZEE® DELUXE EDITION GAME & © 2001 Hasbro, Inc. Used with permission; 25 © FPG.

Design

Luke Malone

www.ingramcontent.com/pod-product-compliance
Lightning Source LLC
Chambersburg PA
CBHW042059210326
41597CB00045B/87